Table of Contents

CHAPTER 1 3
Wired for Success - Your DIY Home Electrical Journey Begins Here 3
CHAPTER 2 4
Fundamentals of Home Electrical Systems 4
   Introduction 4
   Introducing the Importance of Understanding Basic Home Electrical Systems 4
      Brief Overview of Electrical Safety Precautions and Necessary Tools 5
      Key Electrical Components in Homes 7
         Service Entrance 7
         Main Service Panel 7
         Circuit Breakers and Fuses 7
         Wiring 8
      Outlets and Switches 8
   Basics of Electricity 8
      Voltage 8
      Current 9
      Resistance 9
      Circuits 9
How to Safely Turn Off Power to a Circuit or the Entire House.10
   How to Safely Turn Off Power to a Circuit or the Entire House 10
CHAPTER 2 12

Essential Tools and Safety Gear for DIY Electrical Work 12
   Essential Hand Tools for Electrical Work 12
   Safety Gear Recommendations for DIY Electricians 13
   Guidelines for Choosing Quality Electrical Components and Materials 15
Chapter 3 18
DIY Home Electrical Projects 18
   Replacing Electrical Outlets 19
   Installing Light Switches 20
   Upgrading Electrical Panels: Boost Your Home's Power and Safety 21
   Ceiling Fan Installation: A Cool Project for Comfort and Style 23
   Wiring a Home Network: The Ultimate Guide to a Connected Lifestyle 26
   Installing Smart Home Devices: The Future of Home Automation 28
CHAPTER 4 30
Introduction: Advanced DIY Electrical Projects 30
   Installing Ceiling Fans for Optimal Air Circulation 31
   Upgrading Your Home's Lighting with LED Retrofit Kits 31
   Installing Outdoor Lighting for Safety and Aesthetics 31
   Setting Up a Home Solar Power System for Sustainable Energy 32
   Integrating a Whole-House Surge Protector for Enhanced Electrical Safety 32
Chapter 5 34
Troubleshooting and Maintaining Your Home Electrical System 34
   Identifying Common Electrical Issues 34
   Troubleshooting Electrical Problems 35

Preventative Maintenance for Your Electrical System 36
Upgrading and Modernizing Your Electrical System 38
Working with a Licensed Electrician 39
Chapter 6 41
Troubleshooting Common Electrical Issues 41
Circuit Breaker Tripping 41
Dealing with Dead Outlets 43
Troubleshooting Circuit Breaker Trips 44
Identifying and Fixing Hot or Buzzing Outlets 45
Diagnosing Flickering or Dimming Lights 47

CHAPTER 1

Wired for Success - Your DIY Home Electrical Journey Begins Here

Welcome to "Wired for Success: A Comprehensive Guide to DIY Home Electrical Projects"! This book is designed to help you master the fundamentals of home electrical systems and confidently embark on a wide range of DIY projects. From troubleshooting common electrical issues to installing light fixtures, outlets, and switches, our step-by-step instructions will guide you through every aspect of home electrical work. Moreover, we will provide essential information on safety precautions, tools, and best practices to ensure your DIY journey is both enjoyable and successful.

In today's increasingly connected world, a basic understanding of home electrical systems is invaluable for homeowners. Our aim is to demystify these systems and empower you to take control of your home's electrical needs, save money, and enhance your living space. We will also introduce you to the world of smart home technology, enabling you to create a more efficient and convenient home environment.

As you progress through this book, remember that safety is paramount when working with electricity. Always adhere to the recommended safety guidelines and know your limitations. Some projects may still require the expertise of a licensed electrician.

So, let us power up your DIY home electrical journey with "Wired for Success"!

CHAPTER 2
Fundamentals of Home Electrical Systems

Introduction

Welcome to Chapter 1 of "Wired for Success: A Comprehensive Guide to DIY Home Electrical Projects." In this chapter, we will explore the fundamentals of home electrical systems. By understanding the key components and principles that underlie these systems, you will be better equipped to take on DIY projects, troubleshoot issues, and make informed decisions about upgrades and maintenance.

An essential starting point for any DIY electrical journey, this chapter will provide you with a solid foundation upon which to build your skills and knowledge. As you progress through the book, you will be able to refer back to this chapter to reinforce your understanding of the basics and ensure your projects are safe and successful.

Introducing the Importance of Understanding Basic Home Electrical Systems

A solid understanding of basic home electrical systems is the foundation of any successful DIY electrical journey. When you are knowledgeable about the key components and principles of your home's electrical system, you are better equipped to tackle projects, troubleshoot issues, and make informed decisions about upgrades and maintenance.

Knowledge of home electrical systems allows you to:
- Identify potential hazards and take necessary precautions to prevent accidents or damage.
- Enhance the safety, efficiency, and functionality of your living space.
- Save money by handling simple projects yourself and making informed decisions about hiring professionals or purchasing materials.
- Communicate effectively with electricians, ensuring that you get the best value for your money and that the work meets your expectations.

Furthermore, understanding the basics of home electrical systems can be empowering. It can boost your confidence as a homeowner, giving you a sense of control over your living environment. It also allows you to be proactive about maintaining and improving your home's electrical system, which can contribute to a safer and more enjoyable living space.

In this book, we will provide you with the essential knowledge and tools to master the basics of home electrical systems and embark on a wide range of DIY projects with confidence and skill.

Brief Overview of Electrical Safety Precautions and Necessary Tools

Before diving into the world of DIY home electrical projects, it is crucial to familiarize yourself with some fundamental safety precautions and the essential tools you will need. This section provides a brief overview to get you started but remember that more detailed information and guidelines will be provided throughout the book as you tackle specific projects.

Electrical Safety Precautions:

1. Turn off power: Before working on any electrical project, always turn off the power to the circuit or the entire house to avoid electric shock.
2. Test for voltage: Use a voltage tester to confirm that there is no current flowing through the wires you will be working with.
3. Inspect your tools and equipment: Regularly inspect your tools and equipment for wear and tear or damage and replace them as needed.
4. Use insulated tools: Whenever possible, use tools with insulated handles to minimize the risk of electric shock.
5. Wear appropriate safety gear: Protective gloves, safety goggles, and non-conductive footwear can help reduce the risk of injury while working on electrical projects.
6. Keep your workspace dry: Water and electricity are a dangerous combination. Ensure your workspace is dry and free of standing water.
7. Know your limits: Understand your capabilities and recognize when it is best to call a professional electrician for help.

Necessary Tools:
8. Multimeter: This versatile instrument is used to measure voltage, current, and resistance, helping you diagnose and troubleshoot electrical issues.
9. Wire stripper: An essential tool for removing insulation from wires without damaging the conductor.
10. Needle-nose pliers: Useful for bending, cutting, and manipulating wires in tight spaces.
11. Screwdrivers: A variety of screwdrivers, including flat-head and Phillips, are needed for working with several types of screws and electrical components.
12. Voltage tester: This tool helps you confirm whether a wire or device is live, ensuring that it is safe to work with.
13. Wire nuts and electrical tape: Used for connecting and insulating wires safely.
14. Circuit breaker finder: Helps you identify the correct circuit breaker for the circuit you are working on, making it easy to shut off power before starting a project.

By keeping these safety precautions in mind and gathering the necessary tools, you will be well-prepared to tackle a wide range of DIY home electrical projects with confidence and skill.

Key Electrical Components in Homes

To embark on your DIY electrical journey, it is essential to become familiar with the key electrical components found in homes. In this section, we will provide an in-depth analysis of each major component, explaining their purpose and function within the home electrical system.

Service Entrance

The service entrance is the point where electricity enters your home from the utility company. It typically includes the service drop (overhead wires) or service lateral (underground wires), the electric meter, and the service head or weatherhead. The electric meter measures your home's electricity consumption, while the weatherhead provides a weather-resistant entry point for the service drop or service lateral.

Main Service Panel

The main service panel, also known as the breaker box or fuse box, is the central distribution point for electricity in your home. It receives power from the service entrance and distributes it to individual circuits that supply electricity to various parts of your home. The main service panel houses circuit breakers or fuses that protect your home's wiring and electrical devices from overloading and overheating.

Circuit Breakers and Fuses

Circuit breakers and fuses are protective devices designed to interrupt the flow of electricity in case of a fault, such as an overload or short circuit. They prevent damage to your home's wiring and electrical devices, as well as reduce the risk of electrical fires. Circuit breakers can be reset once the issue has been resolved, while fuses need to be replaced.

Wiring

Wiring is the conductive material, typically made of copper or aluminum, that transports electricity throughout your home. Wiring is protected by insulation, which prevents contact with other wires and reduces the risk of electric shock. There are three main types of wires found in residential electrical systems: hot wires (carry current to devices), neutral wires (return current to the service panel), and ground wires (provide a safe path for electricity in case of a fault).

Outlets and Switches

Outlets and switches are devices that control the flow of electricity to appliances, lights, and other electrical devices in your home. Outlets, also known as receptacles, provide a connection point for electrical devices to draw power. There are several types of outlets designed for various purposes, such as standard outlets, GFCI (ground fault circuit interrupter) outlets, and AFCI (arc fault circuit interrupter) outlets.

- Switches, on the other hand, control the flow of electricity to specific devices or circuits. They can be used to turn lights on and off, control fans, or operate other electrical devices. Switches come in a range of styles, such as single-pole, three-way, and dimmer switches, each designed for specific applications.
- Understanding these key electrical components will enable you to confidently approach DIY electrical projects, troubleshoot issues, and effectively communicate with professional electricians when needed.

Basics of Electricity

To effectively work on DIY home electrical projects, it is essential to grasp the fundamental principles of electricity. In this section, we will provide an in-depth analysis of the basics of electricity, including voltage, current, resistance, and circuits.

Voltage

Voltage, also known as electric potential, is the force that pushes electric current through a conductor. It is measured in volts (V) and is often compared to water pressure in a pipe. A higher voltage means a greater force pushing the electric current, while a lower voltage means a weaker force. In residential electrical systems, the standard voltage is typically 120V or 240V, depending on the specific application.

Current

Current is the flow of electric charge, or the movement of electrons, through a conductor. ("Current Electricity: Introduction, Static Electricity, Solved Examples") It is measured in amperes (A), commonly referred to as amps. Like water flowing through a pipe, electric current flows from a higher voltage (or potential) to a lower voltage. In most residential electrical systems, current flows from the service panel, through wiring and devices, and back to the service panel.

Resistance

Resistance is the opposition to the flow of electric current within a conductor or electrical device. It is measured in ohms (Ω) and is often compared to the friction encountered by water flowing through a pipe. Factors that influence resistance include the material of the conductor, its length and diameter, and its temperature. In electrical systems, resistance generates heat, which can be a concern when designing circuits and selecting proper wire sizes.

Circuits

Circuits are closed loops that allow electricity to flow from a power source, through a load (such as a light bulb or an appliance), and back to the source. In residential electrical systems, circuits are typically protected by circuit breakers or fuses within the main service panel. "There are two main types of circuits: series and parallel." ("Difference between Series and Parallel Circuits")

- Series circuits: In a series circuit, devices are connected end-to-end, so there is only one path for current to flow. If one device in the circuit fails or is disconnected, the entire circuit is interrupted, and all devices in the circuit cease to function.
- Parallel circuits: In a parallel circuit, devices are connected side-by-side, creating multiple paths for current to flow. If one device in the

circuit fails or is disconnected, the other devices in the circuit can continue to function.

Understanding the basics of electricity, including voltage, current, resistance, and circuits, is crucial for safely and effectively working on DIY home electrical projects. With this foundational knowledge, you will be better prepared to tackle a wide range of electrical tasks, troubleshoot issues, and make informed decisions about your home's electrical system.

How to Safely Turn Off Power to a Circuit or the Entire House

- Locating your home's main service panel
- Identifying the correct circuit breaker or fuse for the circuit you are working on.
- Turning off the circuit breaker or removing the fuse to disconnect power.
- Using a voltage tester to confirm that power has been successfully turned off.

By mastering the fundamentals of home electrical systems presented in this chapter, you will be well-prepared to tackle a wide range of DIY projects with confidence and skill. As you continue through the book, you will build upon this foundational knowledge to enhance your understanding of home electrical systems and carry out more complex projects safely and effectively.

How to Safely Turn Off Power to a Circuit or the Entire House

Working safely with electricity is crucial, and one of the most important aspects of electrical safety is knowing how to properly turn off power to a circuit or the entire house. In this section, we will provide an in-depth analysis of the steps required to ensure your safety when working on electrical projects.

Locating Your Home's Main Service Panel, the main service panel, also known as the breaker box or fuse box, is the central hub for your home's electrical system. It is typically located in a utility room, basement, or garage. In some homes, it might be located outside. The main service panel contains circuit breakers or fuses that control the flow of electricity to individual circuits throughout your home.

Identifying the Correct Circuit Breaker or Fuse Before working on any electrical project, it is essential to identify the correct circuit breaker or fuse for the circuit you will be working on. The breakers or fuses should be labeled, indicating which areas of your home they control. If they are not labeled, you can use a circuit breaker finder or consult your home's electrical blueprint to determine the correct breaker or fuse.

Turning Off the Circuit Breaker or Removing the Fuse Once you have identified the correct circuit breaker or fuse, turn off the circuit breaker by flipping the switch to the "off" position, or remove the fuse by unscrewing it from its socket. If you need to turn off power to the entire house, locate the main circuit breaker (usually a large, double-pole breaker) and switch it to the "off" position.

Using a Voltage Tester to Confirm Power is Off After turning off the circuit breaker or removing the fuse, it is essential to confirm that there is no electricity flowing through the wires you will be working with. Use a voltage tester, also known as a non-contact voltage tester or a voltage detector, to check for the presence of electrical current. Simply touch the tip of the tester to the wire or device in question. If the tester lights up or beeps, electricity is still present, and you should double-check the circuit breaker or fuse. If the tester does not indicate voltage, it is safe to proceed with your electrical work.

By following these steps to safely turn off power to a circuit or your entire house, you can minimize the risk of electric shock and other hazards while working on DIY home electrical projects. Always exercise caution when working with electricity, and if you are unsure or uncomfortable with any aspect of an electrical project, consult a professional electrician for assistance.

CHAPTER 2

Essential Tools and Safety Gear for DIY Electrical Work

Introduction

Embarking on DIY electrical work requires not only a solid understanding of electrical principles and components but also the right tools and safety gear. In this chapter, we'll discuss the essential tools and safety equipment you'll need to complete various electrical projects, from simple tasks like replacing an outlet to more complex projects like installing new wiring. We'll also provide tips on selecting the best tools and gear for your specific needs, ensuring that you're well-equipped for any electrical task you decide to tackle.

Essential Hand Tools for Electrical Work

Before diving into specific electrical tasks, it's important to have a basic set of hand tools that will be useful for a wide range of projects. These tools are designed to help you manipulate wires, fasten components, and perform other necessary tasks with ease and precision.

Wire Strippers Wire strippers are designed to remove the insulation from wires, allowing you to make proper connections. They come in a variety of styles, including manual and automatic, with different cutting and stripping capacities. Choose a wire stripper that is suitable for the wire gauges you will be working with.

Needle-nose Pliers Needle-nose pliers are useful for gripping, bending, and cutting wires in tight spaces. They have long, narrow jaws that can reach into small areas and provide precise control. Some needle-nose pliers also have built-in wire cutters, making them a versatile tool for various tasks.

Linesman Pliers Linesman pliers, also known as combination pliers, are essential for cutting, gripping, and twisting wires. They have a sturdy design with serrated jaws and a built-in wire cutter. Linesman pliers are perfect for working with heavier-gauge wires and can also be used to cut screws, nails, and other fasteners.

Screwdrivers A good set of screwdrivers is essential for any DIY electrical work. You'll need both flathead and Phillips screwdrivers in various sizes to accommodate different types of screws. Insulated screwdrivers are a safer option, as they protect against accidental contact with live wires.

Voltage Tester A voltage tester, also known as a non-contact voltage tester or voltage detector, is a crucial safety tool for detecting the presence of electrical current in wires and devices. This tool helps ensure that power has been properly turned off before you begin working on a circuit. Choose a voltage tester that's suitable for your home's voltage range and is easy to use.

Electrical Tape Electrical tape is used to insulate and protect electrical connections, such as wire splices and terminal connections. It's made from a non-conductive material, typically vinyl, and is available in a variety of colors to help you identify wires and keep your work organized.

Wire Connectors Wire connectors, also known as wire nuts, are used to join two or more wires together securely. They come in various sizes and styles to accommodate different wire gauges and connection types. Make sure you have a variety of wire connectors on hand to suit your specific project needs.

Cable Ripper A cable ripper is a handy tool for removing the outer sheathing from non-metallic (NM) electrical cable, exposing the individual wires inside. This tool makes it easy to prepare NM cable for wiring projects without damaging the wires.

Fish Tape Fish tape is a flexible, coiled metal tape used to pull wires through conduit or other enclosed spaces. It's an invaluable tool for running new wires in existing walls or ceilings, saving time and effort during your electrical projects.

Safety Gear Recommendations for DIY Electricians

Safety should always be the top priority when working on electrical projects. To minimize the risk of injury, it's essential to have the proper safety gear on hand. In this section, we'll provide a detailed analysis of the recommended safety equipment for DIY electricians, ensuring that you're well-protected while tackling various electrical tasks.

.... Insulated Gloves Insulated gloves are designed to protect your hands from electrical shocks and burns. They're made from non-conductive materials like rubber, which provides an effective barrier against electricity. When selecting insulated gloves, look for those that meet or exceed the International Electrotechnical Commission (IEC) standards for electricians' gloves. Also, make sure to choose gloves that fit well and are comfortable to wear, as you'll be using them for extended periods.

Safety Glasses Safety glasses are crucial for protecting your eyes from flying debris, sparks, and other hazards while working on electrical projects. They should have side shields to provide maximum protection and should meet the American National Standards Institute (ANSI) Z87.1 standards for impact resistance. Choose safety glasses that fit comfortably and have anti-fog and scratch-resistant lenses for clear visibility.

Face Shield A face shield is an essential safety gear for tasks that involve cutting, grinding, or drilling, as these activities can generate sparks and flying debris. The face shield provides full-face protection, ensuring that your eyes, nose, and mouth are shielded from potential hazards. Look for a face shield with an adjustable headband and a clear, impact-resistant visor for optimal protection and comfort.

Arc Flash Protection Arc flashes are sudden releases of electrical energy that can cause severe burns and other injuries. If you're working on projects that involve high-voltage systems or potential arc flash hazards, it's essential to wear arc flash protective clothing. This can include flame-resistant coveralls, jackets, and pants, as well as arc-rated face shields and hoods. Make sure to select arc flash protective gear that meets the appropriate ASTM and NFPA 70E standards.

Safety Boots Safety boots provide protection for your feet while working on electrical projects. They should have non-conductive, slip-resistant soles to prevent electrical shocks and reduce the risk of falls. In addition, safety boots with steel or composite toe caps can help protect your feet from impact and compression injuries. When selecting safety boots, look for those that meet or exceed the ASTM F2412-18 and F2413-18 standards for electrical hazard protection.

Hearing Protection Some electrical tasks, such as drilling or cutting, can generate loud noises that can damage your hearing. Wearing earplugs or earmuffs can help protect your ears from excessive noise levels. Choose hearing protection that's comfortable to wear and provides adequate noise reduction for your specific needs.

Hard Hat A hard hat is essential safety gear when working in areas where there's a risk of falling objects, such as construction sites or when working above head height. Hard hats provide protection for your head from impact injuries and should meet the ANSI Z89.1 standards for industrial head protection. Choose a hard hat with an adjustable suspension system for a secure, comfortable fit.

By equipping yourself with the proper safety gear, you can significantly reduce the risk of injury while working on DIY electrical projects. Always prioritize your safety, and if you're unsure or uncomfortable with any aspect of an electrical task, consult a professional electrician for assistance.

Guidelines for Choosing Quality Electrical Components and Materials

Using quality electrical components and materials is essential for ensuring the safety, efficiency, and longevity of your DIY electrical projects. Poor quality components can lead to malfunctions, increased energy consumption, and potential hazards. In this section, we'll provide detailed guidelines for selecting the best electrical components and materials to guarantee the success of your projects.

Compliance with Standards and Codes Ensure that the components you purchase are compliant with industry standards, such as the National Electrical Code (NEC) in the United States, or the IEC (International Electrotechnical Commission) standards for international use. Compliance with these codes and standards is an indication of the quality and safety of electrical components.

Look for UL, ETL, or CSA Listings When purchasing electrical components, look for products that are UL (Underwriters Laboratories), ETL (Intertek), or CSA (Canadian Standards Association) listed. These marks indicate that the product has undergone rigorous testing and meets stringent safety and performance standards.

Choose Reputable Brands and Manufacturers Select components from reputable brands and manufacturers with a proven track record of producing high-quality products. Research customer reviews and expert recommendations to identify the best brands in the industry. Investing in high-quality components may cost more upfront but can save you money and hassle in the long run.

Understand the Different Types of Wire Choosing the right type of wire for your project is crucial for safety and performance. Be familiar with the various types of wire available, such as non-metallic (NM) cable, armored cable (AC or BX), and metal-clad (MC) cable. Each type has specific uses and is suitable for different applications. Always consult the NEC or your local electrical code to ensure you're using the correct type of wire for your project.

Consider Wire Gauge Wire gauge, or the thickness of the wire, is another crucial factor when selecting electrical components. Thicker wires have lower gauge numbers and can carry more current than thinner wires. Ensure that you choose the appropriate wire gauge for the electrical load of your project, as specified by the NEC or your local electrical code.

Select High-Quality Connectors and Terminals Using high-quality connectors and terminals is essential for ensuring secure and reliable connections. Look for connectors made from corrosion-resistant materials like brass or copper, which offer better conductivity and durability. Additionally, choose terminals that are compatible with the wire gauge and type you're using.

Invest in Durable, Weather-Resistant Materials If your project involves outdoor or damp locations, select electrical components and materials that are designed for these environments. Look for components that are rated for wet or damp locations, such as weather-resistant receptacles and switches, and use conduit or cable suitable for outdoor use.

Read Product Specifications and Labels Always read product specifications and labels to understand the limitations and capabilities of the components you're purchasing. Pay attention to factors such as voltage and amperage ratings, temperature ranges, and compatibility with other components or systems. This information will help you make informed decisions and ensure the success of your electrical projects.

By following these guidelines, you can choose quality electrical components and materials that will help you complete safe, efficient, and long-lasting DIY electrical projects. Remember to prioritize safety and consult with a professional electrician if you're unsure about any aspect of your project.

Chapter 3
DIY Home Electrical Projects

Electrical projects are a common aspect of homeownership and can range from simple tasks like replacing outlets and switches to more complex projects like upgrading electrical panels or installing home automation systems. As a homeowner, having the knowledge and skills to tackle some of these projects yourself can be empowering, cost-effective, and even enjoyable. In this chapter, we will delve into several DIY home electrical projects that can help you improve your living space, enhance your home's safety, and increase its overall value.

We will begin by discussing the basics of replacing electrical outlets, a common and relatively simple project that can make a significant difference to the appearance and functionality of your rooms. We will then move on to installing light switches, a task that can transform the lighting options in your home and enhance its ambiance. Upgrading electrical panels will be our next focus, as we explore the benefits of modernizing your home's electrical infrastructure and ensuring it meets your power needs. Furthermore, we will discuss the installation of ceiling fans, which provide both aesthetic and functional benefits to your living spaces.

As technology continues to advance, home networks have become an essential aspect of modern living, and in this chapter, we will guide you through the process of wiring a home network. In addition, outdoor lighting plays a crucial role in safety and aesthetics, and we will provide a comprehensive guide to installing various types of exterior illumination. Finally, we will delve into the world of home automation and smart devices, helping you navigate the process of integrating cutting-edge technology into your home's electrical system.

By the end of this chapter, you will have gained valuable knowledge and skills that will empower you to tackle various DIY home electrical projects with confidence and ensure the safety and efficiency of your electrical system.

Replacing Electrical Outlets

Replacing electrical outlets is a common DIY home electrical project that can have a significant impact on the appearance and functionality of your living spaces. Worn-out or damaged outlets can be unsightly and pose safety hazards, while upgrading to modern outlets with features like USB charging ports can provide added convenience. In this section, we will discuss the steps for safely replacing an electrical outlet.

1. Turn off the power: Before starting any electrical work, ensure that the power is turned off at the breaker box to prevent the risk of electric shock. Locate the circuit breaker that controls the outlet you'll be working on and switch it off.
2. Confirm the power is off: Use a voltage tester or a non-contact voltage detector to confirm that there is no electricity flowing to the outlet. Simply insert the tester's probes into the outlet slots or hold the detector near the outlet to ensure it's safe to work on.
3. Remove the outlet cover: Unscrew the outlet cover using a screwdriver and gently remove it from the wall.
4. Unscrew the outlet: Next, unscrew the outlet from the electrical box by removing the screws on the top and bottom of the outlet. Carefully pull the outlet out of the box, exposing the wiring connections.
5. Disconnect the wires: Take note of the wiring connections before disconnecting the wires from the old outlet. You may want to take a photo for reference. Loosen the terminal screws and gently remove the wires from the old outlet.
6. Connect the wires to the new outlet: Attach the wires to the new outlet in the same configuration as the old one. Typically, black (hot) wires connect to the brass screws, white (neutral) wires connect to the silver screws, and green or bare copper (ground) wires connect to the green grounding screw.
7. Secure the new outlet: Carefully push the outlet back into the electrical box and screw it into place. Replace the outlet cover and secure it with the screw.
8. Restore power: Turn the power back on.

Installing Light Switches

Installing or replacing light switches is a practical skill that can significantly improve the lighting options and atmosphere of your home. In this section, we will cover the different types of light switches, the necessary tools, and a step-by-step guide for installing light switches.

Types of Light Switches:
15. Single-pole switch: The most common type of light switch, used for controlling a single light or group of lights from one location.
16. Three-way switch: Allows you to control a single light or group of lights from two different locations, typically used for stairways, hallways, or large rooms.
17. Dimmer switch: Provides variable control over the brightness of your lights, allowing you to create different moods and save energy.
18. Smart switch: Offers remote control and programmability through a smartphone or home automation system, giving you greater control over your home's lighting.

Tools and Materials Needed:
- New light switch
- Screwdriver (flathead and Phillips)
- Wire stripper
- Needle-nose pliers
- Voltage tester or non-contact voltage detector
- Wire nuts (if needed)

Step-by-Step Guide to Installing a Light Switch:
1. Turn off the power: Always turn off the power at the breaker box before working on any electrical project. Locate the circuit breaker that controls the light switch you're working on and switch it off.
2. Remove the switch cover: Unscrew the switch cover using a screwdriver and gently remove it from the wall.
3. Verify the power is off: Use a voltage tester or a non-contact voltage detector to confirm that there is no electricity flowing to the switch.
4. Remove the old switch: Unscrew the switch from the electrical box and carefully pull it out to expose the wiring connections. Note or take a picture of the wiring configuration before disconnecting the wires from the old switch.

5. Disconnect the wires: Loosen the terminal screws on the old switch and gently remove the wires.
6. Prepare the new switch: If necessary, use a wire stripper to strip the insulation from the ends of the wires. Then, create a loop at the end of each wire using needle-nose pliers to ensure a secure connection with the new switch.
7. Connect the wires to the new switch: Attach the wires to the new switch in the same configuration as the old one. For single-pole switches, the black (hot) wires typically connect to the brass screws, while the green or bare copper (ground) wire connects to the green grounding screw. For three-way switches, follow the manufacturer's instructions to ensure proper wiring connections.
8. Secure the new switch: Carefully push the switch back into the electrical box and screw it into place. Replace the switch cover and secure it with the screw.
9. Restore power: Turn the power back on at the breaker box and test the new switch to ensure it's working correctly.

By following these steps, you can successfully install light switches in your home and improve the lighting options and atmosphere in your living spaces. Remember to always prioritize safety and consult with a professional electrician if you're unsure about any aspect of your project.

Upgrading Electrical Panels: Boost Your Home's Power and Safety

Upgrading your home's electrical panel is a critical project that can enhance safety, increase the number of circuits, and ensure compatibility with modern appliances. In this section, we will discuss the reasons for upgrading your electrical panel and provide a detailed overview of the process.

Reasons to Upgrade Electrical Panels:
1. Increased capacity: older homes may not have the electrical capacity to support contemporary appliances and electronics. Upgrading to a higher-capacity panel can prevent overloaded circuits and potential hazards.

2. Improved safety: Worn or damaged panels can pose a risk of electrical fires. Upgrading to a modern panel with updated safety features can protect your home and its occupants.
3. Accommodating new appliances: large appliances, such as electric vehicles or central air conditioning systems, may require dedicated circuits. Upgrading your panel can ensure you have enough circuits to support these appliances.

Upgrading Electrical Panels: A Step-by-Step Overview
1. Assess your home's needs: Determine the appropriate panel size and type for your home by evaluating your current and future electrical requirements. Consider your existing appliances, electronics, and any planned additions or renovations.
2. Consult a professional electrician: Upgrading an electrical panel can be complex and potentially dangerous. Consult a professional electrician to review your home's needs and provide guidance on the panel upgrade process.
3. Obtain necessary permits: Check with your local building department to determine if you need a permit for the panel upgrade. Following local regulations can ensure the safety and legality of your project.
4. Turn off the power: Before starting any electrical work, ensure that the power is turned off at the main breaker box.
5. Remove the old panel: With the power off, carefully remove the old panel by disconnecting the circuits and unscrewing the panel from the wall. Be cautious and consult the professional electrician if you are unsure about any steps.
6. Install the new panel: Follow the manufacturer's instructions and guidance from the professional electrician to install the new panel. Ensure proper grounding and secure the panel to the wall.
7. Reconnect the circuits: Carefully reconnect the circuits to the new panel, ensuring that each circuit is correctly labeled and connected to the appropriate breaker.
8. Restore power: Turn the power back on at the main breaker box and test the circuits to ensure they are functioning correctly.

Please note that upgrading an electrical panel is a complex project that requires a thorough understanding of electrical systems and safety precautions. It's essential to consult with a professional electrician to ensure a safe and successful panel upgrade. By upgrading your electrical panel, you can significantly improve your home's safety, capacity, and compatibility with modern appliances.

Ceiling Fan Installation: A Cool Project for Comfort and Style

Installing a ceiling fan is a practical and stylish home improvement project that can enhance the comfort, energy efficiency, and aesthetic appeal of your living spaces. In this section, we will discuss the benefits of installing a ceiling fan and provide a detailed, step-by-step guide to help you complete this DIY electrical project.

Benefits of Installing a Ceiling Fan:
1. 3.4 Ceiling Fan Installation: A Cool Project for Comfort and Style
2. Installing a ceiling fan is a practical and stylish home improvement project that can enhance the comfort, energy efficiency, and aesthetic appeal of your living spaces. In this section, we will discuss the benefits of installing a ceiling fan and provide a detailed, step-by-step guide to help you complete this DIY electrical project.
3. Benefits of Installing a Ceiling Fan:
4. Improved comfort: Ceiling fans circulate air, creating a comfortable breeze that can make your home feel cooler in the summer and warmer in the winter.
5. Energy efficiency: Using a ceiling fan can help reduce your reliance on air conditioning and heating systems, potentially lowering your energy bills.

6. Aesthetic appeal: With a wide range of styles and designs available, ceiling fans can be an attractive addition to your home's decor.
7. Step-by-Step Guide to Installing a Ceiling Fan:
8. Choose the right fan: Select a ceiling fan that suits your room size, ceiling height, and style preferences. Be sure to choose a fan rated for the specific location, such as indoor or outdoor use.
9. Gather tools and materials: You will need a screwdriver, wire stripper, wire nuts, a voltage tester, a ladder, and the ceiling fan kit, which should include the fan, mounting bracket, and hardware.
10. Turn off the power: Before starting any electrical work, ensure that the power is turned off at the breaker box.
11. Remove existing light fixture or fan: If you're replacing an existing fixture, carefully remove it, disconnect the wires, and unscrew the mounting bracket.
12. Install the new mounting bracket: Attach the new mounting bracket to the ceiling electrical box using the provided screws. Ensure it is secure and level.
13. Assemble the fan: Follow the manufacturer's instructions to assemble the fan motor, blades, and any other components. Some fans may come partially assembled, simplifying this step.
14. Wire the fan: With the fan assembly in hand, connect the fan's wiring to the wiring in the ceiling electrical box. Typically, you'll need to connect the white (neutral) wires, black or red (hot) wires, and green or bare copper (ground) wires using wire nuts. Consult the manufacturer's instructions for specific wiring details.
15. Attach the fan to the bracket: Carefully lift the fan assembly and hook it onto the mounting bracket. This may require the assistance of another person to support the fan while you secure it with screws.
16. Install the light kit (if applicable): If your fan includes a light kit, follow the manufacturer's instructions to assemble and install it. Connect the light kit's wiring to the fan's wiring using wire nuts.

17. Install fan blades and trim: Attach the fan blades to the fan motor using the provided screws, followed by any decorative trim or covers.
18. Restore power and test the fan: Turn the power back on at the breaker box and test the fan and light (if applicable) to ensure proper operation.
19. By following this step-by-step guide, you can successfully install a ceiling fan in your home, enhancing the comfort, energy efficiency, and style of your living spaces. Remember to prioritize safety and consult with a professional electrician if you're unsure about any aspect of the project.
20. Improved comfort: Ceiling fans circulate air, creating a comfortable breeze that can make your home feel cooler in the summer and warmer in the winter.
21. Energy efficiency: Using a ceiling fan can help reduce your reliance on air conditioning and heating systems, potentially lowering your energy bills.
22. Aesthetic appeal: With a wide range of styles and designs available, ceiling fans can be an attractive addition to your home's decor.

Step-by-Step Guide to Installing a Ceiling Fan:
1. Choose the right fan: Select a ceiling fan that suits your room size, ceiling height, and style preferences. Be sure to choose a fan rated for the specific location, such as indoor or outdoor use.
2. Gather tools and materials: You will need a screwdriver, wire stripper, wire nuts, a voltage tester, a ladder, and the ceiling fan kit, which should include the fan, mounting bracket, and hardware.
3. Turn off the power: Before starting any electrical work, ensure that the power is turned off at the breaker box.
4. Remove existing light fixture or fan: If you're replacing an existing fixture, carefully remove it, disconnect the wires, and unscrew the mounting bracket.
5. Install the new mounting bracket: Attach the new mounting bracket to the ceiling electrical box using the provided screws. Ensure it is secure and level.
6. Assemble the fan: Follow the manufacturer's instructions to assemble the fan motor, blades, and any other components.

Some fans may come partially assembled, simplifying this step.
7. Wire the fan: With the fan assembly in hand, connect the fan's wiring to the wiring in the ceiling electrical box. Typically, you'll need to connect the white (neutral) wires, black or red (hot) wires, and green or bare copper (ground) wires using wire nuts. Consult the manufacturer's instructions for specific wiring details.
8. Attach the fan to the bracket: Carefully lift the fan assembly and hook it onto the mounting bracket. This may require the assistance of another person to support the fan while you secure it with screws.
9. Install the light kit (if applicable): If your fan includes a light kit, follow the manufacturer's instructions to assemble and install it. Connect the light kit's wiring to the fan's wiring using wire nuts.
10. Install fan blades and trim: Attach the fan blades to the fan motor using the provided screws, followed by any decorative trim or covers.
11. Restore power and test the fan: Turn the power back on at the breaker box and test the fan and light (if applicable) to ensure proper operation.

By following this step-by-step guide, you can successfully install a ceiling fan in your home, enhancing the comfort, energy efficiency, and style of your living spaces. Remember to prioritize safety and consult with a professional electrician if you're unsure about any aspect of the project.

Wiring a Home Network: The Ultimate Guide to a Connected Lifestyle

In today's digital age, a reliable home network is crucial for staying connected to the world and enjoying various online services. A well-designed and properly installed home network can greatly enhance your internet speed, device connectivity, and overall online experience. In this section, we will discuss the benefits of a wired home network and provide a detailed, step-by-step guide to help you design and install your own network.

Benefits of Wiring a Home Network:

1. Improved connectivity: A wired network provides a stable and fast connection for devices such as computers, gaming consoles, and smart TVs, ensuring seamless online experiences.
2. Enhanced security: Wired connections are generally more secure than wireless networks, reducing the risk of unauthorized access or data breaches.
3. Scalability: A wired home network can be easily expanded as your needs grow, allowing for the addition of new devices or services without compromising performance.

Step-by-Step Guide to Wiring a Home Network:
1. Plan your network: Determine the locations of your devices and the most efficient route for running network cables between them. Consider future expansion when planning cable runs and network equipment placement.
2. Choose the right cables: Select high-quality Ethernet cables, such as Cat 6 or Cat 7, to ensure optimal performance and futureproofing. Measure the required cable lengths and purchase additional cable to account for any unforeseen obstacles or adjustments.
3. Gather tools and materials: You will need a drill, wire stripper, cable crimper, Ethernet connectors, wall plates, a patch panel, a network switch, and cable management supplies, such as cable clips or conduits.
4. Install wall plates and outlets: Drill holes and install wall plates at the designated locations for each device. Attach Ethernet connectors to the cable ends and secure them in the wall plates.
5. Run the cables: Route the network cables from each device location to a central point, such as a utility closet or basement. Use cable clips or conduits to secure and organize the cables as needed.
6. Terminate the cables: Strip and terminate the cable ends at the central point using a patch panel. This provides a neat and organized solution for connecting multiple cables to your network switch.
7. Connect devices to the network switch: Using short patch cables, connect each terminated cable on the patch panel to the corresponding port on the network switch.

8. Configure your network: Connect your modem or router to the network switch and configure your network settings, such as assigning static IP addresses or setting up a DHCP server, if necessary.
9. Test your connections: Verify that each device on your network can connect to the internet and communicate with other devices on the network.

By following this step-by-step guide, you can successfully wire a home network that provides reliable and secure connections for all your devices. Remember to prioritize safety and consult with a professional electrician or network specialist if you're unsure about any aspect of the project. With a wired home network in place, you can enjoy a connected lifestyle and make the most of your online experiences.

Installing Smart Home Devices: The Future of Home Automation

Smart home devices are transforming the way we live, offering unprecedented convenience, energy efficiency, and security. By integrating smart devices into your home, you can control lighting, appliances, temperature, and more using voice commands, smartphone apps, or automated routines. In this section, we will discuss the benefits of smart home technology and provide a detailed, step-by-step guide to help you install and set up your own smart home devices.

Benefits of Smart Home Devices:
1. Convenience: Manage your home's devices from anywhere, using voice commands or smartphone apps, making everyday tasks more accessible and efficient.
2. Energy efficiency: Smart home devices can optimize energy usage by learning your habits, automatically adjusting settings, and providing usage insights to help you save on your utility bills.
3. Enhanced security: Smart security devices, such as cameras, doorbells, and locks, can provide real-time monitoring and alerts, increasing the safety of your home and its occupants.

Step-by-Step Guide to Installing Smart Home Devices:

1. Plan your smart home: Determine which devices you want to install and where they will be placed in your home. Consider compatibility with your existing devices and the type of smart home ecosystem you prefer, such as Amazon Alexa, Google Assistant, or Apple HomeKit.
2. Choose the right devices: Select smart home devices that suit your needs and preferences, considering factors such as ease of installation, compatibility, and functionality.
3. Gather tools and materials: Depending on the specific devices you are installing, you may need a screwdriver, drill, wire stripper, or other tools. Review the manufacturer's instructions for each device to determine the required tools and materials.
4. Install and connect the devices: Follow the manufacturer's instructions to install each device. This may involve mounting devices to walls or ceilings, connecting wires, or plugging in power adapters. Ensure that each device is securely installed and properly connected to your home's electrical system.
5. Configure your devices: Connect each device to your home's Wi-Fi network and follow the manufacturer's instructions to set up the device using a smartphone app or web interface. Configure settings such as device names, automation routines, and access permissions, as needed.
6. Integrate with your smart home ecosystem: Link your devices to your preferred smart home ecosystem, such as Amazon Alexa, Google Assistant, or Apple HomeKit, by following the specific instructions for each platform.
7. Test your smart home devices: Ensure that each device is functioning correctly and responds to voice commands, smartphone app controls, or automated routines as expected.

By following this step-by-step guide, you can successfully install and set up a range of smart home devices, transforming your living space into a modern, connected home. Remember to prioritize safety and consult with a professional electrician or smart home specialist if you're unsure about any aspect of the project. With your smart home devices in place, you can enjoy unparalleled convenience, energy efficiency, and security in your daily life.

In conclusion, Chapter 3 has provided you with comprehensive information and step-by-step guides on various DIY electrical projects to enhance your home's safety, efficiency, and functionality. We began by discussing the importance of understanding basic home electrical systems and safety precautions, followed by an in-depth analysis of key electrical components and the basics of electricity.

Throughout this chapter, we have explored essential tools and safety gear for DIY electrical work, guidelines for choosing quality electrical components, and tips for installing and maintaining various devices. We have covered practical projects like replacing outlets, switches, and light fixtures, as well as more advanced topics like wiring a home network and installing smart home devices.

Additionally, we have shared troubleshooting tips for common electrical issues and guidelines on when to call a professional electrician. The knowledge and skills acquired in this chapter will empower you to undertake various electrical tasks confidently and safely while saving money on repairs and upgrades.

As you continue to explore the world of DIY electrical projects, always prioritize safety and seek professional assistance when needed. With practice and patience, you'll become more proficient in handling electrical tasks, resulting in a safer, more efficient, and enjoyable living environment.

CHAPTER 4
Introduction: Advanced DIY Electrical Projects

Welcome to Chapter 4, where we will dive into advanced DIY electrical projects that can further improve your home's safety, efficiency, and aesthetics. Building on the foundational knowledge acquired in the previous chapters, we will explore more complex tasks, such as installing ceiling fans, upgrading to LED lighting, setting up outdoor lighting, integrating a home solar power system, and installing a whole-house surge protector. By following the detailed guides provided in this chapter, you can confidently undertake these projects while prioritizing safety and efficiency. Let's begin our journey into advanced DIY electrical projects!

Installing Ceiling Fans for Optimal Air Circulation

Ceiling fans offer numerous benefits, including improved air circulation, energy efficiency, and added aesthetic appeal to your home. To install a ceiling fan, follow these steps:
1. Select the appropriate fan size and mounting option based on your room's dimensions and ceiling height.
2. Turn off the power at the circuit breaker and remove the existing light fixture or electrical box.
3. Install a fan-rated electrical box and secure the fan bracket to the box using the provided screws.
4. Assemble the fan motor and canopy, then hang the motor assembly on the bracket.
5. Wire the fan by connecting the appropriate wires and securing them with wire nuts.
6. Attach the fan blades, light kit (if applicable), and any additional accessories.
7. Test the fan's operation by restoring power and using the provided remote or wall switch.

Upgrading Your Home's Lighting with LED Retrofit Kits

LED retrofit kits offer enhanced energy efficiency and improved lighting quality. To upgrade your home's lighting with LED retrofit kits, follow these steps:
8. Determine the type of retrofit kit needed based on your existing fixtures and lighting preferences.
9. Turn off the power at the circuit breaker and remove the existing lighting fixture.
10. Install the LED retrofit kit, following the manufacturer's instructions for mounting and wiring.
11. Secure the new LED components in place, restore power, and test the new lighting.

Installing Outdoor Lighting for Safety and Aesthetics

Outdoor lighting provides increased safety, enhanced curb appeal, and extended outdoor living space usability. To install outdoor lighting, follow these steps:

1. Choose the right types of outdoor lighting fixtures based on your needs and preferences.
2. Plan your outdoor lighting layout, considering factors such as pathway illumination, accent lighting, and security lighting.
3. Install the fixtures, ensuring they are securely mounted and properly wired.
4. Connect the outdoor lighting to your home's electrical system and test the functionality.

Setting Up a Home Solar Power System for Sustainable Energy

A home solar power system offers reduced energy costs, environmental sustainability, and increased property value. To set up a home solar power system, follow these steps:

1. Size and select the right solar power system components, including solar panels, inverters, and mounting hardware.
2. Install the solar panels on your roof or a ground-mounted array, following local regulations and guidelines.
3. Mount the inverter and connect it to the solar panels, your home's electrical system, and the utility grid.
4. Test the system's operation and monitor its performance using the provided monitoring tools.

Integrating a Whole-House Surge Protector for Enhanced Electrical Safety

A whole-house surge protector safeguards your home's electrical system and devices from power surges. To install a whole-house surge protector, follow these steps:

1. Choose the right surge protector based on your home's electrical system and protection needs.
2. Turn off the power at the circuit breaker and install the surge protector in your electrical panel, following the manufacturer's instructions for grounding and connecting to the panel's circuit breakers.
3. Test the surge protector's operation by restoring power.

By following these detailed guides for Sections 4.1 through 4.5, you can confidently tackle advanced DIY electrical projects to enhance your home's safety, efficiency, and aesthetics. Always prioritize safety and consult with a professional electrician when needed. These projects will not only save you money but also contribute to a more sustainable and enjoyable living environment.

Chapter 5
Troubleshooting and Maintaining Your Home Electrical System

In Chapter 5, we will delve into the essential aspects of troubleshooting and maintaining your home electrical system. A well-maintained electrical system ensures safety, efficiency, and longevity. This chapter will provide you with the knowledge and practical guidance needed to diagnose common electrical issues and perform routine maintenance tasks. By following the detailed instructions provided in this chapter, you'll be better equipped to handle electrical problems and keep your home's electrical system in top condition.

Identifying Common Electrical Issues

In this section, we'll explore some of the most common electrical issues homeowners may encounter and help you understand the root causes of these problems. By identifying these issues early, you can take appropriate action to maintain the safety and efficiency of your electrical system.

1. Flickering Lights: Flickering lights can be a nuisance and may indicate a loose or faulty connection in the circuit. This issue can often be resolved by tightening the light bulb or replacing it if it's defective. However, if the problem persists, it might indicate a more severe issue with the wiring or connections, which requires further investigation.
2. Tripped Circuit Breakers: Circuit breakers are designed to trip when an electrical circuit becomes overloaded or experiences a short circuit. If you find that a particular circuit breaker trips frequently, it may be a sign of an overloaded circuit or a faulty breaker. To resolve the issue, try

redistributing the electrical load or consult a professional electrician to diagnose and repair the problem.
3. Malfunctioning Outlets: If an electrical outlet stops working or appears to be loose, it may be due to a worn-out or damaged receptacle. In some cases, replacing the receptacle can resolve the issue. However, if you notice scorch marks or other signs of damage around the outlet, it's essential to consult an electrician as it could indicate a more serious problem with your home's wiring.
4. Warm or Buzzing Switches and Outlets: If you notice that a switch or outlet is warm to the touch or emits a buzzing sound, it could be a sign of a loose connection or an overloaded circuit. These issues can pose a fire risk, so it's crucial to address them promptly by consulting a professional electrician.
5. High Electricity Bills: If you notice a sudden spike in your electricity bills, it could be due to inefficient appliances, outdated wiring, or faulty connections. Conduct an energy audit to identify the sources of energy waste and implement energy-saving measures to reduce your overall consumption.

By being aware of these common electrical issues and understanding their root causes, you can take proactive steps to maintain the safety and efficiency of your home's electrical system. In many cases, identifying and addressing these problems early can prevent more significant issues from developing and ensure the longevity of your electrical system.

Troubleshooting Electrical Problems

In this section, we'll provide step-by-step guidance on how to troubleshoot various electrical issues you may encounter in your home. By following these practical tips and methods, you can diagnose and address problems with circuit breakers, wiring, outlets, and more.
1. Circuit Breakers: If a circuit breaker trips frequently, first determine whether the circuit is overloaded by unplugging some devices and resetting the breaker. If the problem persists, the breaker itself may be faulty and need replacement. When in doubt, consult a professional electrician.

2. Wiring Issues: If you suspect problems with your home's wiring, such as flickering lights or warm outlets, first turn off the power at the main circuit breaker. Inspect visible wiring for signs of wear, damage, or improper connections. If you identify any issues, contact a licensed electrician to perform repairs or replace the wiring.
3. Outlets: For malfunctioning outlets, first check whether a circuit breaker has tripped, or a fuse has blown. If the problem isn't resolved, turn off the power at the circuit breaker and use a voltage tester to ensure the outlet is not live. Remove the outlet's cover plate and inspect the wiring connections. Tighten any loose connections or replace the outlet if it's damaged or worn out. If you're unsure about any of these steps, seek the assistance of a professional electrician.
4. Light Fixtures: If a light fixture isn't working, first try replacing the bulb with a new one. If the issue persists, turn off the power at the circuit breaker and use a voltage tester to confirm that the fixture is not live. Inspect the wiring connections and secure any loose connections. If the light fixture still doesn't work, it may be faulty and require replacement.
5. Dimming or Flickering Lights: If lights are dimming or flickering, first try tightening or replacing the bulbs. If the problem continues, it could be an issue with the fixture, wiring, or electrical panel. Consult a professional electrician to diagnose and repair the issue.
6. Warm or Buzzing Switches and Outlets: If a switch or outlet feels warm or buzzes, turn off the power at the circuit breaker and inspect the connections. Tighten any loose connections, replace damaged components, or consult an electrician if you're uncertain about the cause or solution.

By following these troubleshooting steps, you can diagnose and address many common electrical issues in your home. Always prioritize safety by turning off the power before working with electrical components and consult with a professional electrician when needed. By taking a proactive approach to troubleshooting electrical problems, you can maintain the safety and efficiency of your home's electrical system.

Preventative Maintenance for Your Electrical System

In this section, we will focus on the importance of regular maintenance to ensure the safety and efficiency of your home's electrical system. By performing routine maintenance tasks, you can keep your electrical system in top condition and avoid potential hazards.

1. Inspect Wiring: Regularly inspect visible wiring in your home for signs of wear, damage, or improper connections. Look for frayed or cracked wires, loose connections, and any signs of overheating, such as scorch marks or discoloration. If you find any issues, consult a licensed electrician to perform the necessary repairs or replace the wiring.
2. Test Outlets: Use an outlet tester to check the functionality and safety of your electrical outlets. The tester will indicate whether the outlet is wired correctly and if there are any potential issues, such as reversed polarity or an open ground. Address any problems by tightening connections, replacing damaged receptacles, or consulting a professional electrician.
3. Check Circuit Breakers: Regularly inspect your electrical panel for signs of wear or damage, such as rust or water leakage. Test circuit breakers by turning them off and on to ensure they are functioning correctly. If you encounter a breaker that trips frequently or fails to reset, consult a professional electrician to diagnose and repair the issue.
4. Maintain GFCI Outlets: Ground Fault Circuit Interrupter (GFCI) outlets are essential for protecting you from electrical shocks in areas where water is present, such as bathrooms, kitchens, and outdoor spaces. Test your GFCI outlets monthly by pressing the "test" button on the outlet and ensuring that the power is cut off. If the outlet fails to reset, replace it or consult a professional electrician.
5. Clean Light Fixtures and Replace Bulbs: Regularly clean light fixtures to remove dust and debris, which can cause overheating and reduce the lifespan of your bulbs. Replace burnt-out or flickering bulbs with energy-efficient LED bulbs to improve efficiency and reduce electricity consumption.
6. Schedule Regular Electrical Inspections: To ensure the safety and efficiency of your electrical system, schedule regular electrical inspections with a licensed electrician. These inspections can help identify potential issues before they become significant problems, saving you time, money, and frustration in the long run.

By performing these preventative maintenance tasks, you can keep your home's electrical system in top condition, ensuring safety, efficiency, and longevity. Being proactive with maintenance not only prevents potential hazards but also helps you identify and address issues before they escalate, ultimately saving you time and money.

Upgrading and Modernizing Your Electrical System

While many DIY electrical projects can be carried out safely by homeowners with basic knowledge and skills, certain situations require the expertise of a professional electrician. In this section, we will discuss the circumstances in which you should consult a professional and the benefits of doing so.

1. Complex Electrical Projects: For projects that involve extensive rewiring, the installation of new circuits, or working directly with the electrical panel, it's crucial to call a licensed electrician. These tasks can be hazardous if not performed correctly and may also require permits and inspections to ensure compliance with local building codes.
2. Persistent Issues: If you've attempted to troubleshoot an electrical issue using the methods described in this book but haven't resolved the problem, it's time to consult a professional. A licensed electrician can quickly diagnose the issue and provide an effective solution, saving you time and frustration.
3. Signs of Overheating or Damage: If you notice scorch marks, discoloration, or other signs of overheating or damage around your electrical components, it's essential to call a professional electrician immediately. These issues can pose serious fire risks and need to be addressed promptly.
4. Upgrading Your Electrical System: If you're planning to renovate your home or add new appliances that require additional electrical capacity, consult an electrician to assess your current system and recommend any necessary upgrades. This ensures that your system can safely handle the increased load and meets all relevant safety standards.
5. Inadequate Knowledge or Experience: If you're unsure about how to approach an electrical project or feel uncomfortable

working with electrical components, it's always best to call a professional. A licensed electrician has the necessary training, experience, and equipment to perform the job safely and efficiently.

By knowing when to call a professional electrician, you can ensure the safety and efficiency of your home's electrical system. In many cases, enlisting the help of a professional not only guarantees that the job is done correctly but can also save you time, money, and potential hazards. Remember, safety should always be your top priority when working with electricity.

Working with a Licensed Electrician

In this section, we will discuss the significance of having regular electrical inspections performed by a licensed professional electrician. These inspections can help identify potential issues, maintain safety, and ensure the efficiency of your home's electrical system.

1. Safety: The primary reason for regular electrical inspections is to ensure the safety of your home and its occupants. A professional electrician can identify and address any potential hazards, such as faulty wiring, overloaded circuits, or damaged electrical components, before they escalate into dangerous situations.
2. Compliance with Codes and Regulations: Electrical inspections can help ensure that your home's electrical system complies with local building codes and safety regulations. A professional electrician will be familiar with these requirements and can recommend any necessary upgrades or repairs to keep your home in compliance.
3. Energy Efficiency: Regular electrical inspections can help improve the energy efficiency of your home. A professional electrician can identify areas where energy is being wasted, such as outdated lighting or inefficient appliances, and recommend upgrades to help reduce energy consumption and lower your utility bills.

4. Preventative Maintenance: By identifying potential issues early, regular electrical inspections can prevent the need for costly repairs or replacements in the future. An electrician can spot and fix minor problems before they turn into significant issues, saving you time and money in the long run.
5. Home Value: A well-maintained electrical system can help maintain or even increase the value of your home. If you're planning to sell your home, a recent electrical inspection can provide prospective buyers with added confidence in the safety and functionality of the electrical system.
6. Insurance Requirements: Some insurance providers may require regular electrical inspections as a condition of your home insurance policy. By scheduling these inspections, you can ensure that your policy remains valid and avoid potential issues with claims.

By scheduling regular electrical inspections, you can maintain the safety, efficiency, and value of your home while also ensuring compliance with codes and regulations. A professional electrician can identify and address potential issues, provide recommendations for upgrades and improvements, and help you maintain a safe and efficient electrical system for years to come.

By the end of Chapter 5, you will have a solid understanding of how to troubleshoot and maintain your home's electrical system. Armed with this knowledge, you can ensure the safety, efficiency, and longevity of your electrical system, making your home a more comfortable and enjoyable place to live.

Chapter 6

Troubleshooting Common Electrical Issues

In this chapter, we will explore common electrical issues that homeowners may encounter, along with their possible causes and solutions. By learning to recognize and troubleshoot these problems, you can save time, money, and potentially prevent more significant issues from developing.

Chapter 6 Introduction: Troubleshooting Common Electrical Issues

As a homeowner, encountering electrical issues is almost inevitable. While some problems may require the expertise of a professional electrician, many common issues can be resolved with a basic understanding of electrical systems and some troubleshooting skills. In this chapter, we will discuss several frequently encountered electrical problems, their potential causes, and the solutions you can implement to restore your home's electrical system to proper working order. By learning to recognize and address these issues, you can save time, money, and ensure the safety and efficiency of your electrical system.

Circuit Breaker Tripping

Circuit breakers are designed to protect your home's electrical system by tripping and interrupting the electrical flow when an issue arises. While it is normal for a circuit breaker to trip occasionally, frequent tripping can indicate a more significant problem. In this section, we will explore the common causes of circuit breaker tripping and offer solutions to address these issues.

1. Overloaded Circuits: An overloaded circuit occurs when the electrical demand on a circuit exceeds its capacity. This can be caused by having too many appliances or devices running

simultaneously. To resolve this issue, try redistributing the electrical load by unplugging some devices or moving them to different circuits.
2. Short Circuits: A short circuit happens when a live wire comes into contact with a neutral or ground wire, causing a surge in current. This can result from damaged wiring, improper connections, or faulty appliances. To troubleshoot a short circuit, start by turning off the affected circuit and inspecting the wiring and connected devices for visible damage. If you cannot identify the source of the problem, it is best to consult a professional electrician.
3. Ground Faults: A ground fault occurs when a live wire comes into contact with a grounded object, such as a metal conduit or a grounded appliance. This can pose a serious safety risk, as it may cause electrical shocks or fires. To address a ground fault, inspect the affected circuit for damaged wiring or malfunctioning appliances. If the issue persists, contact a licensed electrician for further assistance.

By understanding the common causes of circuit breaker tripping, you can more effectively troubleshoot the problem and take appropriate action. Remember that while some issues can be resolved through simple adjustments, more complex problems may require the expertise of a professional electrician to ensure the safety and functionality of your home's electrical system.

6.2 Flickering or Dimming Lights

Flickering or dimming lights can be both annoying and potentially indicative of an underlying electrical issue. In this section, we will discuss the possible causes of flickering or dimming lights and provide guidance on how to address these problems.
1. Loose Connections: One of the most common reasons for flickering or dimming lights is a loose connection, either at the light fixture or within the electrical wiring. To check for loose connections, first, turn off the power to the affected light fixture at the circuit breaker. Then, inspect the connections at the light fixture and ensure they are tight and

secure. If the problem persists, it may be due to a loose connection in the wiring, in which case a professional electrician should be consulted.
2. Overloaded Circuits: If several lights on the same circuit are flickering or dimming, it could be a sign of an overloaded circuit. In this case, try reducing the electrical load on the circuit by unplugging some devices or switching them to a different circuit. If the problem continues, consult an electrician to determine if a circuit upgrade is necessary.
3. Faulty Light Bulbs: Sometimes, the issue may be as simple as a faulty light bulb. To rule this out, try replacing the flickering bulb with a new one. If the new bulb flickers as well, the problem likely lies elsewhere in the electrical system.
4. Voltage Fluctuations: Flickering or dimming lights can also be caused by voltage fluctuations in your home's electrical system. These fluctuations can result from large appliances, such as air conditioners or refrigerators, cycling on and off. If you suspect voltage fluctuations are causing your lights to flicker or dim, consider having a professional electrician perform a voltage test to determine if any corrective action is needed.

By understanding the possible causes of flickering or dimming lights, you can more effectively troubleshoot the issue and take the necessary steps to resolve it. Keep in mind that while some problems can be fixed with simple adjustments, others may require the expertise of a professional electrician to ensure your home's electrical system remains safe and functional.

Dealing with Dead Outlets

A dead outlet can be a frustrating inconvenience, but it is also a sign that there might be an issue with your home's electrical system. In this section, we will explore the potential causes of dead outlets and provide guidance on how to address these issues safely and effectively.
1. Tripped Circuit Breaker or GFCI: Sometimes, a dead outlet is the result of a tripped circuit breaker or GFCI (ground fault circuit interrupter) outlet. To check if this is the cause, locate

your home's electrical panel and inspect the circuit breakers for any that have been tripped. If you find a tripped breaker, reset it by flipping it off and then back on. If the outlet is still dead, check for nearby GFCI outlets and press the reset button to restore power.
2. Loose or Damaged Wiring: If the issue is not related to a tripped breaker or GFCI, it could be caused by loose or damaged wiring. First, turn off the power to the affected outlet at the circuit breaker. Then, carefully remove the outlet cover and inspect the wiring connections. If any wires are loose or disconnected, reattach them securely. If the wiring appears damaged or frayed, contact a professional electrician for further assistance.
3. Worn or Faulty Outlets: Over time, outlets can become worn or damaged, leading to a loss of power. To determine if the outlet is faulty, turn off the power at the circuit breaker, and use a voltage tester to confirm there is no electricity flowing to the outlet. Next, carefully remove the outlet and inspect it for signs of wear or damage. If the outlet appears to be in poor condition, replace it with a new one.
4. Unknown Causes: If you have exhausted the above troubleshooting steps and the outlet remains dead, it is time to call in a professional electrician. They can perform a more thorough inspection of your home's electrical system to identify and address the underlying issue.

By understanding the potential causes of dead outlets and following the appropriate troubleshooting steps, you can safely and effectively address these issues. Always exercise caution when working with electricity, and do not hesitate to consult a professional electrician if you are unsure how to proceed.

Troubleshooting Circuit Breaker Trips

Circuit breakers are designed to trip when they detect an electrical fault, such as an overload or short circuit. While this safety mechanism is essential for protecting your home's electrical system, it can be frustrating when a breaker trips frequently. In this section, we will discuss the common causes of circuit breaker trips and provide guidance on how to address these issues.

1. Overloaded Circuits: An overloaded circuit occurs when the combined electrical demand of the devices connected to the circuit exceeds its capacity. To resolve this issue, first identify the devices that are causing the overload, then either unplug some of them or redistribute them to other circuits. If the problem persists, consult a professional electrician to determine if a circuit upgrade is necessary.
2. Short Circuits: A short circuit happens when a hot wire comes into contact with a neutral or ground wire, causing a sudden surge of electricity that trips the breaker. To identify a short circuit, turn off the power to the affected circuit at the electrical panel, then inspect the wiring and outlets for signs of damage or burning. If you find evidence of a short circuit, contact a professional electrician to address the issue.
3. Ground Faults: Ground faults are similar to short circuits but occur when a hot wire comes into contact with a grounded object or surface. Ground faults are particularly dangerous in areas with moisture, such as bathrooms and kitchens, which is why GFCI outlets are required in these locations. If you suspect a ground fault is causing your breaker to trip, consult a professional electrician for assistance.
4. Faulty Circuit Breakers: In some cases, the issue may lie with the circuit breaker itself. Over time, breakers can become worn or damaged, leading to frequent trips. If you have ruled out other causes and suspect a faulty breaker, consult a professional electrician to replace the breaker.

By understanding the common causes of circuit breaker trips and taking the appropriate troubleshooting steps, you can safely and effectively address these issues. Remember, working with electricity can be dangerous, so always exercise caution and consult a professional electrician if you are unsure how to proceed.

Identifying and Fixing Hot or Buzzing Outlets

A hot or buzzing outlet can be a warning sign of a potential electrical problem in your home. In this section, we will discuss the

possible causes of hot or buzzing outlets and provide guidance on how to safely address these issues.
1. Overloaded Outlets: One of the primary reasons for a hot or buzzing outlet is an overload caused by too many devices plugged into the outlet or the use of high-wattage appliances. To resolve this issue, unplug some of the devices or redistribute them to other outlets. If the problem persists, consider upgrading the outlet to one with a higher amperage rating or consult a professional electrician for further assistance.
2. Loose or Damaged Wiring: Hot or buzzing outlets can also be the result of loose or damaged wiring connections. To check for this issue, first, turn off the power to the affected outlet at the circuit breaker. Next, carefully remove the outlet cover and inspect the wiring connections. If any wires are loose, disconnected, or damaged, contact a professional electrician to address the issue.
3. Worn or Faulty Outlets: Over time, outlets can become worn or damaged, which may cause them to overheat or produce a buzzing sound. If you suspect the outlet is faulty, first turn off the power at the circuit breaker and use a voltage tester to confirm there is no electricity flowing to the outlet. Then, carefully remove the outlet and inspect it for signs of wear or damage. If necessary, replace the outlet with a new one.
4. Arcing: Arcing occurs when an electrical current jumps across a gap between conductors, generating heat and potentially causing a fire. If you hear a buzzing or crackling sound coming from an outlet, it could be a sign of arcing. In this case, contact a professional electrician immediately to assess the situation and make any necessary repairs.

By understanding the possible causes of hot or buzzing outlets and taking the appropriate steps to address these issues, you can help ensure the safety and efficiency of your home's electrical system. Always exercise caution when working with electricity and consult a professional electrician if you are unsure how to proceed.

Diagnosing Flickering or Dimming Lights

Flickering or dimming lights can be more than just an annoyance; they can indicate an underlying electrical issue in your home. In this section, we will discuss the possible causes of flickering or dimming lights and provide guidance on how to safely address these issues.

1. Loose or Damaged Light Bulbs: The most straightforward cause of flickering or dimming lights is a loose or damaged light bulb. To check if this is the issue, turn off the power to the affected light fixture, and ensure the light bulb is properly screwed in. If the bulb appears damaged, replace it with a new one.
2. Fluctuating Voltage: If multiple lights in your home are flickering or dimming, it could be due to fluctuating voltage. This issue can be caused by a variety of factors, such as large appliances starting up or a problem with your utility company's electrical supply. To diagnose and address voltage fluctuations, contact a professional electrician for assistance.
3. Overloaded Circuits: Flickering or dimming lights can also be a symptom of an overloaded circuit. If you suspect this is the issue, try redistributing the electrical load by unplugging some devices or moving them to other circuits. If the problem persists, consult a professional electrician to determine if a circuit upgrade is necessary.
4. Loose or Damaged Wiring: Flickering or dimming lights can also be caused by loose or damaged wiring connections. To check for this issue, first, turn off the power to the affected light fixture at the circuit breaker. Then, carefully remove the fixture and inspect the wiring connections. If any wires are loose or damaged, contact a professional electrician to address the issue.
5. Faulty Light Switches: A faulty light switch can cause flickering or dimming lights. To determine if the switch is the issue, turn off the power to the affected light fixture at the circuit breaker, and use a voltage tester to confirm there is no electricity flowing to the switch. Next, carefully remove the switch and inspect it for signs of wear or damage. If necessary, replace the switch with a new one.

By understanding the possible causes of flickering or dimming lights and taking the appropriate steps to address these issues, you can help ensure the safety and efficiency of your home's electrical system. Always exercise caution when working with electricity and consult a professional electrician if you are unsure how to proceed.

By understanding the potential causes and solutions for common electrical issues, you'll be better equipped to troubleshoot problems and take appropriate actions. By addressing these issues promptly, you can maintain the safety and efficiency of your home's electrical system and prevent more significant problems from developing.

In conclusion, "Home Wiring Unplugged: A DIY Guide to Simple Electrical Projects" has provided you with a comprehensive understanding of home electrical systems, safety precautions, tools, and practical knowledge needed to perform a variety of DIY electrical projects.

Throughout the book, we have discussed essential topics such as understanding the basics of electricity, key electrical components in homes, and safety gear recommendations for DIY electricians. Additionally, we have explored the importance of choosing quality electrical components and materials to ensure the safety and functionality of your projects.

Our step-by-step guide has covered various projects, from simple tasks like replacing outlets and switches to more advanced projects like installing new light fixtures and diagnosing common electrical issues. By now, you should feel empowered with the knowledge and skills to tackle these projects confidently and safely.

Moreover, we have emphasized the importance of understanding the limitations of DIY electrical work and when to consult a professional electrician. This awareness will help you make informed decisions, prioritize safety, and avoid potential hazards.

As you embark on your DIY electrical journey, always remember to follow the safety precautions outlined in this book and seek professional help when needed. With the knowledge you have gained from "Home Wiring Unplugged," you will be better equipped to enhance the safety, efficiency, and functionality of your home while saving money on repairs and upgrades.

We hope you found this guide informative and valuable, and we encourage you to continue learning and expanding your DIY electrical skills. Good luck with your future projects and remember—safety first!